EARLY BIRD STORIES™

City Animals

An Alien's Guide

Alex Francis

Early★Reader

Lerner Publications ◆ Minneapolis

First American edition published in 2020 by Lerner Publishing Group, Inc.

Illustrated by Alex Francis

First published by Maverick Arts Publishing Limited

Licensed Edition
City Animals: An Alien's Guide

Lerner Publications Company
An imprint of Lerner Publishing Group, Inc.
241 First Avenue North
Minneapolis, MN 55401 USA

For reading levels and more information, look up this title at www.lernerbooks.com.

Main body text set in Mikado. Typeface provided by HVD Fonts.

Library of Congress Cataloging-in-Publication Data
The Cataloging-in-Publication Data for *City Animals: An Alien's Guide* is on file at the Library of Congress.
ISBN 978-1-72841-505-5 (lib. bdg.)
ISBN 978-1-72841-521-5 (pbk)
ISBN 978-1-72841-522-2 (eb pdf)

Manufacured in the United States of America
1-48661-49084-1/15/2020

Table of Contents

Hello fellow aliens!
What do you want to know
about planet Earth?

INCOMING MESSAGE

Dear Finn and Zeek,

We want to visit Earth. We're hoping to visit some cities but want to see animals as well.

Can you tell us about some of the animals that live in cities?

Sincerely,
Tik and Tok
Planet Sit-ee

Introduction

Many types of animals live in cities. One reason is that cities are getting larger. As cities grow, they take over spaces where animals live. So animals have to start living in the cities too.

Living in the City

Animals that live in the city have lots of problems.

- They can't move around as freely as they would in the countryside.
- They may not be able to get the food they normally eat.
- They may be in danger from traffic.
- Some animals get sick because of **pollution.**
- They may not get along well with humans.

But there are some good things about living in the city.

- Animals can eat food waste left behind by humans.
- Animals living in cities may not be hunted by as many **predators.**

Pigeons

Pigeons live everywhere in cities! The city is a good place for pigeons because there are lots of tall buildings. Pigeons build their nests on the tall buildings.

Peregrine Falcons

Peregrine falcons are hunting birds. They like to hunt smaller birds like pigeons. So when pigeons moved to the city, peregrine falcons moved there too.

Foxes

Foxes have been living in cities for at least fifty years!

In fact, there are often more foxes in the city than in the countryside.

Foxes like living in cities because there is lots of food for them to eat, from bugs to rats to human garbage. Foxes are smart, so they have quickly learned how to deal with the dangers of the city.

Leopards

In the big city of Mumbai in India, you might see leopards! Leopards live alongside people in this big city. The city is a good place for the leopards because there are lots of wild dogs. Leopards hunt and eat the wild dogs.

Wild dogs are dangerous because they have **diseases**—so the leopards are helping people by hunting the dogs!

House Geckos

House geckos live in Southeast Asia. You can tell from their name that house geckos live in houses.

The lights in houses **attract** bugs, and the geckos eat the bugs!

Tanuki

The biggest city in Japan is Tokyo. Lots of people live in Tokyo, and lots of animals live there too.

One animal is the tanuki (called a racoon dog in English). Some tanuki even live on the grounds of the Imperial Palace!

Rowdy Neighbors

Wild Boars

Wild boars have moved into German towns and cities. They usually come to the city in the winter months when they are looking for food. Wild boars are shy, but they dig up the green parts of the city. They also leave big piles of mud. This is annoying for the people who live in the city.

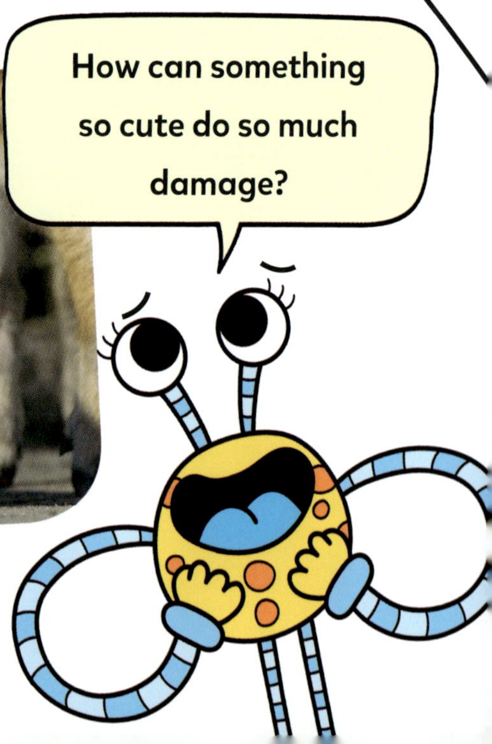

Boars dig up the grass because they are looking for food. They don't eat garbage. They like eating acorns!

Chacma Baboons

Sometimes animals and humans don't mix very well. Chacma baboons cause lots of trouble in Cape Town, South Africa. They often break into houses, take over restaurants, and even steal food right out of people's hands!

Uh oh . . .

Black Bears

Black bears have moved into some cities in North America. They like eating garbage from people's trash bins.

Bears are very smart. They know they will find food where humans live. They sometimes break into cars and houses to steal food.

Polar Bears

Polar bears are shy, and they don't want to live in cities. But because the **polar ice caps** have been melting, some polar bears have had to move to cities to find food.

Hungry polar bears are dangerous to humans. So humans are trying to move polar bears back to safer places, away from cities.

MESSAGE SENT

Dear Tik and Tok,

Cities can be scary places for animals. But some animals like city life! Humans are helping some animals find new homes outside the city.

If you are visiting Japan, be sure to visit the sika deer in Nara. They live in the city and can be fed special deer snacks!

From,
Finn and Zeek

Sika deer in Japan

Quiz

1. Which of these is not so much of a danger for animals in cities?
    a) Pollution
    b) Traffic
    c) Sunburn

2. Where can you find city leopards?
    a) Mumbai, India
    b) London, United Kingdom
    c) Tokyo, Japan

3. What is a tanuki called in English?
    a) Fox dog
    b) Raccoon dog
    c) Dog bear

4. What do boars eat?
   a) Acorns
   b) Garbage
   c) Cars

5. Which of these animals doesn't mix well with humans?
   a) House geckos
   b) Pigeons
   c) Chacma baboons

6. Polar bears go into cities for food because . . . ?
   a) They want to eat garbage.
   b) They like the lights.
   c) The ice caps are melting, and they have to find food somewhere else.

Quiz Answers:
1c, 2a, 3b, 4a, 5c, 6c

Glossary

attract: to feel a pull toward something

diseases: illnesses that can hurt people, animals, or plants

polar ice caps: big areas of ice at the North Pole and South Pole

pollution: harmful waste that can hurt the environment

predators: animals that hunt other animals

Index

COLOR	GRL
Silver	L-P
Gold	K-L
Purple	J-K
Orange	H-J
Green	G-I
Blue	E-G
Yellow	C-E
Red	C-D
Pink	A-C

EARLY BIRD STORIES™

Leveled for Guided Reading

Early Bird Stories have been edited and leveled by leading educational consultants to correspond with guided reading levels. The levels are assigned by taking into account the content, language style, layout, and phonics used in each book.